INCREDIBLE TECHNOLOGY

by Meg Marquardt

12 STORY LIBRARY

www.12StoryLibrary.com

12-Story Library is an imprint of Bookstaves and Press Room Editions

Produced for 12-Story Library by Red Line Editorial

Photographs ©: LaurentVu/Sipa/AP Images, cover, 1; Stoyan Totov/Shutterstock Images, 4; Pasquale Sorrentino/Science Source, 5; 10 FACE/Shutterstock Images, 6; welcomia/Shutterstock Images, 7; Andrea Izzotti/Shutterstock Images, 8; Fat Jackey/Shutterstock Images, 9; Everett Historical/Shutterstock Images, 10; Chen feibo/Imaginechina/AP Images, 11; Darryl Dyck/The Canadian Press/AP Images, 12; Chitose Suzuki/AP Images, 13; Wally Santana/AP Images, 14; Tinxi/Shutterstock Images, 15, 28; Dimarion/Shutterstock Images, 16; Oleksandr Rupeta/NurPhoto/Sipa/AP Images, 17; Lee Prince/Shutterstock Images, 18; lightpoet/Shutterstock Images, 19; Daniel Jedzura/Shutterstock Images, 20; Paul Brown/Rex Features/AP Images, 21; Kyodo/AP Images, 22, 29; goodmoments/Shutterstock Images, 23; VanderWolf Images/Shutterstock Images, 24; Sigur/Shutterstock Images, 25; bbernard/Shutterstock Images, 26; NASA/Goddard/Chris Gunn, 27

Content Consultant: Sunil S. Mehendale, assistant professor, School of Technology, Michigan Tech

Library of Congress Cataloging-in-Publication Data
Names: Marquardt, Meg, author.
Title: Incredible technology / Meg Marquardt.
Description: Mankato, MN : 12 Story Library, 2017. | Series: Unbelievable |
    Includes bibliographical references and index. | Audience: Grades 4 to 6.
Identifiers: LCCN 2016047405 (print) | LCCN 2016048182 (ebook) | ISBN
    9781632354228 (hardcover : alk. paper) | ISBN 9781632354921 (pbk. : alk.
    paper) | ISBN 9781621435440 (hosted e-book)
Subjects: LCSH: Technological innovations--Juvenile literature.
Classification: LCC T173.8 .M3665 2017 (print) | LCC T173.8 (ebook) | DDC
    600--dc23
LC record available at https://lccn.loc.gov/2016047405

Printed in the United States of America
022017

Access free, up-to-date content on this topic plus a full digital version of this book. Scan the QR code on page 31 or use your school's login at 12StoryLibrary.com.

Table of Contents

Augmented Reality Could Re-create Ancient Worlds

Augmented reality is taking the world by storm. This incredible technology blends the real world with computer-generated images. It is changing how people interact with the world, even the ancient one.

Augmented reality has been used mostly for games. In the summer of 2016, *Pokémon GO* brought video games into the real world. Phone in hand, players captured computer-generated Pokémon in the real world. The game used the cameras on smartphones to scan users' surroundings. Pokémon appeared on top of images in the real world. Players captured Pokémon on their front porches or in their refrigerators.

A *Pokémon GO* player detects a Drowzee in July 2016.

Augmented reality helps visitors to ruins in Pompeii, Italy, virtually travel back in time.

Groups of Pokémon showed up in public places. Players followed.

The incredible technology is still very new. But the future of augmented reality is boundless. One promising use is the reconstruction of ruins. Archaeologists spend their whole careers investigating ancient sites.

They study the ruins of structures that have collapsed long ago. What they find helps them understand what those structures may have looked like. Augmented reality could take that understanding a step further. Archaeologists could create interactive images of dig sites. Augmented reality could help them see what the site looked like long ago. It could help them make an educated guess as to how the site was used.

1 billion
Estimated augmented reality users worldwide by 2020.

- Augmented reality combines computer-generated images and the real world.
- It can be used to make interactive games.
- It can also be used to re-create long-lost buildings and civilizations.

THINK ABOUT IT

What other uses for augmented reality can you think of? How might augmented reality help doctors? Engineers? Scientists? Use your imagination to describe how augmented reality might be used in the future.

Cryptocurrency Changes the Face of Money

Imagine traveling from the United States to Germany. A US dollar bill cannot be used to purchase things in Germany. Germany uses another type of currency called the Euro. American dollars must be exchanged for Euros. Then they can be used in shops in Germany and most other European countries. But travelers cannot exchange physical bills or coins for cryptocurrency. It exists entirely online.

Cryptocurrency is digital money. It is not like the money in a bank account or wallet. There are no coins or bills to store or exchange for goods. Cryptocurrency is used exclusively to buy, sell, and trade things online.

This digital currency has several advantages. Cryptocurrency cannot be imitated or tampered with the way physical money can be. It is also extremely secure. This is a benefit for online transactions. Cryptocurrency is not exchanged with other currencies, as US dollars and Euros are. This can be helpful for people who live and work in different countries. They do not have

People buy and sell bitcoins online on the bitcoin website.

$672.27
Average cost, in US dollars, of a single bitcoin in October 2016.

- Cryptocurrency is digital money.
- It is often used to buy and sell goods online.
- Cryptocurrency is secure, but its price can change dramatically.

THINK ABOUT IT

What are some of the pros and cons of using cryptocurrency? Would you ever use it? Why or why not?

to worry about constantly exchanging currencies.

But cryptocurrency has some drawbacks. Cryptocurrency is not regulated by a government. This allows the value of cryptocurrency to change, sometimes dramatically. A cryptocurrency called the bitcoin usually costs a few hundred dollars per unit to buy. But at times, it has soared to more than $1,000 per coin. Though cryptocurrency is secure, it is possible to lose it. Technical flaws such as a crashed hard drive can erase digital wallets. Hackers could break into digital bitcoin accounts. The lost cryptocurrency cannot be recovered.

Cryptocurrency that hackers steal cannot be recovered.

Desalination Plants Make Salty Water Drinkable

Water covers approximately 70 percent of the world's surface. But most of this water is undrinkable. Only approximately 1 percent of Earth's water is safe to drink. Scientists call drinkable water potable water. This 1 percent is not enough to support the world's growing population. Researchers must develop technologies that make more water potable.

One promising technology is desalination. Desalination removes salt from water. This process can make ocean water drinkable. It can remove salt from salty groundwater.

Desalination plants follow a process. Pumps pull up water from a salty source, such as the ocean.

A desalination plant in Hamburg Harbor, Hamburg, Germany

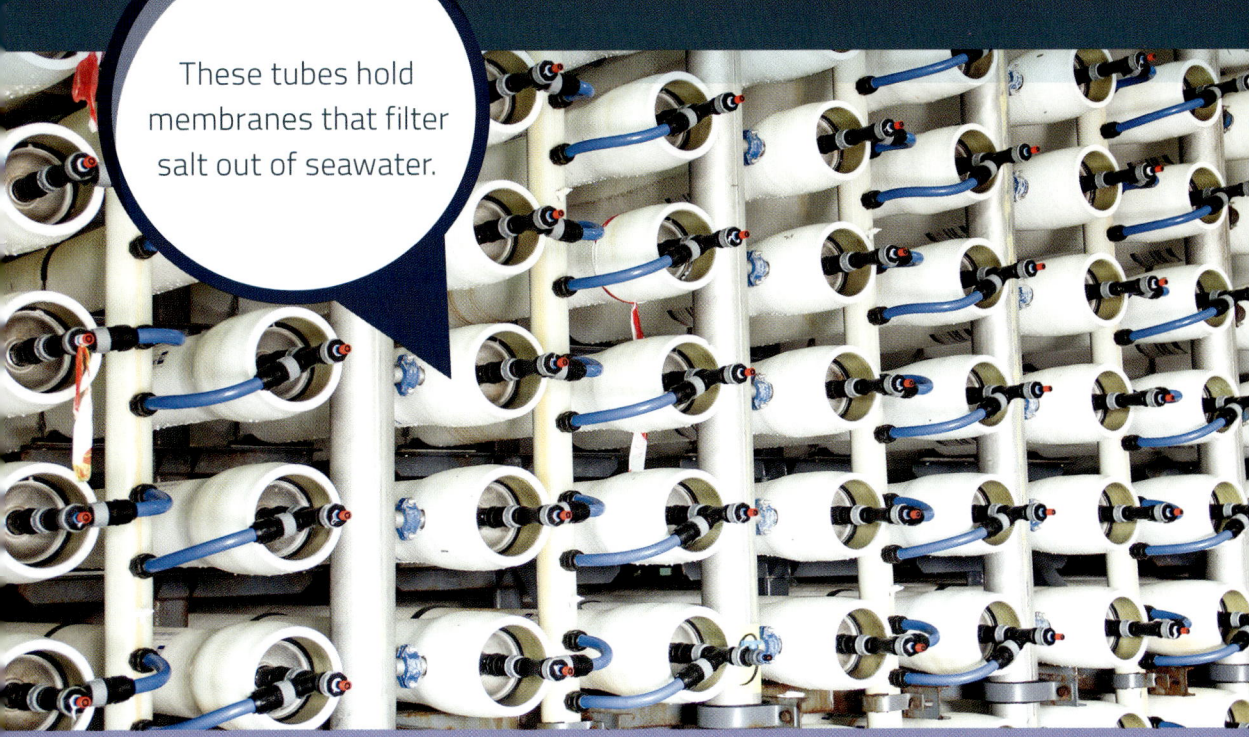

These tubes hold membranes that filter salt out of seawater.

That water is pushed through a membrane. The membrane is a thin layer of plastic. It has holes so small only water molecules can pass through it. The bigger salt molecules are left behind. Every 2 gallons (7.6 L) of salt water create 1 gallon (3.8 L) of potable water.

18,426

Number of desalination plants worldwide in 2015.

- Desalination plants make salty water drinkable.
- The plants push water through a membrane, leaving the salt behind.
- Desalination plants use a lot of energy and can pollute surrounding seawater.

Desalination plants could increase the world's drinkable water supply. But there are still many hurdles. The plants use up to 10 times as much energy as traditional drinking water treatment plants do. Leftover salt often gets dumped back into the ocean. This can make the surrounding area too salty for sea life to survive. Future desalination technologies could fix these problems.

Drones Deliver to Your Doorstep

Modern drone technology has existed for more than 20 years. These small, unmanned aircraft have many uses. Militaries use drones in wars. This incredible technology can watch the enemy on the ground from high in the sky. Moviemakers use drones to shoot movie scenes. But soon, drones could have an entirely different use. They could deliver the mail or even a pizza.

It takes a long time for a truck to deliver a package. Often, roads are not the most direct route. A truck may travel far out of the way to deliver a package. But drones would not have to follow the roads, as trucks do. A drone could take the shortest distance between two points. This would cut down on delivery time.

A US military predator drone takes off from an air base in 2008.

5

Weight, in pounds (2.3 kg), of a package a drone could deliver.

- Drones have many valuable military and commercial uses.
- In the future, drones might deliver packages.
- Lightweight drones can carry only small packages.

DELIVERING MEDICAL AID

Drones could be used to deliver more than just packages. They may one day deliver vital medical aid. A drone could deliver medicine or even blood to a war-torn region. Doctors would not have to take risky trips to restock supplies. Drones could be especially helpful in regions where no medical help is available.

But drone delivery has its drawbacks. Drones cannot carry very heavy loads. This would limit the types of packages drones carry. There is also the problem of range. Most drones run on battery power.

This reduces air pollution. But most drones have a range of only 15 miles (24 km). With current technology, drone delivery would be good for only short trips.

A Chinese retailer tests drone delivery in November 2016.

11

Electronic Contact Lens Sees Farther Than Ever

The next wave of wearable technology could tell users more than the number of steps they take. Electronic contact lenses could change the way people interact in and with the world. A user could take a picture in a literal blink of an eye. These lenses could enable a user to see in the dark. They could even monitor a user's heart rate or blood sugar levels.

Electronic contacts are a very new technology. Scientists are working on different designs. One design is similar to a typical contact lens. A wearer could put a lens in and take it out with ease. But other designs are more permanent. Lenses would be surgically inserted into the eye. The lens would stay in place forever. Or at least until the wearer wanted an upgrade.

Engineering such tiny electronics is tricky. Researchers must develop a way to power the electronic lenses, especially implanted ones.

This permanent lens could replace a pair of eyeglasses.

Like the mesh, this brain implant could connect the brain to a computer.

They have dreamed up many cool solutions to this problem. Lenses could be solar powered. Or they could be powered by a blink. Every time a user blinks, the movement creates energy. That energy could be used to power electronic lenses.

1
Thickness, in micrometers, of the circuitry in an electronic contact lens.

- Electronic contact lenses could be helpful in many different ways.
- They can be used to monitor medical information or help users see in the dark.
- Some electronic contact lenses are removable, but future lenses could be permanently implanted.

EMBEDDED CIRCUITRY

Researchers have not limited themselves to electronic contact lenses. Some scientists have developed a brain implant. The implant is a circuit. It looks like metal mesh. A computer can talk directly to the circuit. Each wire of the field talks to an individual brain cell. A doctor can use the computer to control the brain. This technology could be helpful for people suffering from brain injuries.

Food Technology Feeds a Growing Population

By 2050, the world population will be close to 10 billion. With that many mouths to feed, the pressure on the food industry will be higher than ever. Luckily, two new food technologies may help keep everyone full.

Indoor farming may one day help feed millions of people. Traditional outdoor farming relies on sunlight, nutrient-rich soil, and ideal weather. A growing season that is too rainy or too dry can cause crops to fail. But indoor farming does not require ideal weather. It does not require sunlight or even soil. Instead, farmers use LED lights that create perfect growing conditions. An indoor farm could take up a single room or thousands of square feet in a huge warehouse.

Fruits and vegetables grown indoors do not need soil. Instead, some farmers use a technique called hydroponics. Plants sit in a soil-like substance that allows water to pass through it efficiently. Nutrient-rich water is continually passed over

An indoor, vertical hydroponics farm in Taiwan

PRINTING FOOD

The future of food technology could be the 3-D printer. The "ink" of a 3-D printer can be made of edible substances. These substances can be used to create a meal. Future 3-D printer ink may be made of insect protein or algae. The printer would use this edible ink to build 3-D foods. These bountiful protein sources could someday feed the world.

40

Number of plants, per square foot (0.09 sq m), that can be grown in an indoor farm.

- New food technologies are needed to feed the world's growing population.
- Indoor farmers grow food without sunlight, rainfall, or soil.
- Farmers use hydroponics or aeroponics to grow indoor crops.

the plants' roots. Other farmers use a technique called aeroponics. Farmers suspend plants on metal grates. They spray a nutrient-rich

mist on the plants' roots. Indoor farming is an incredible technology that could help feed the world.

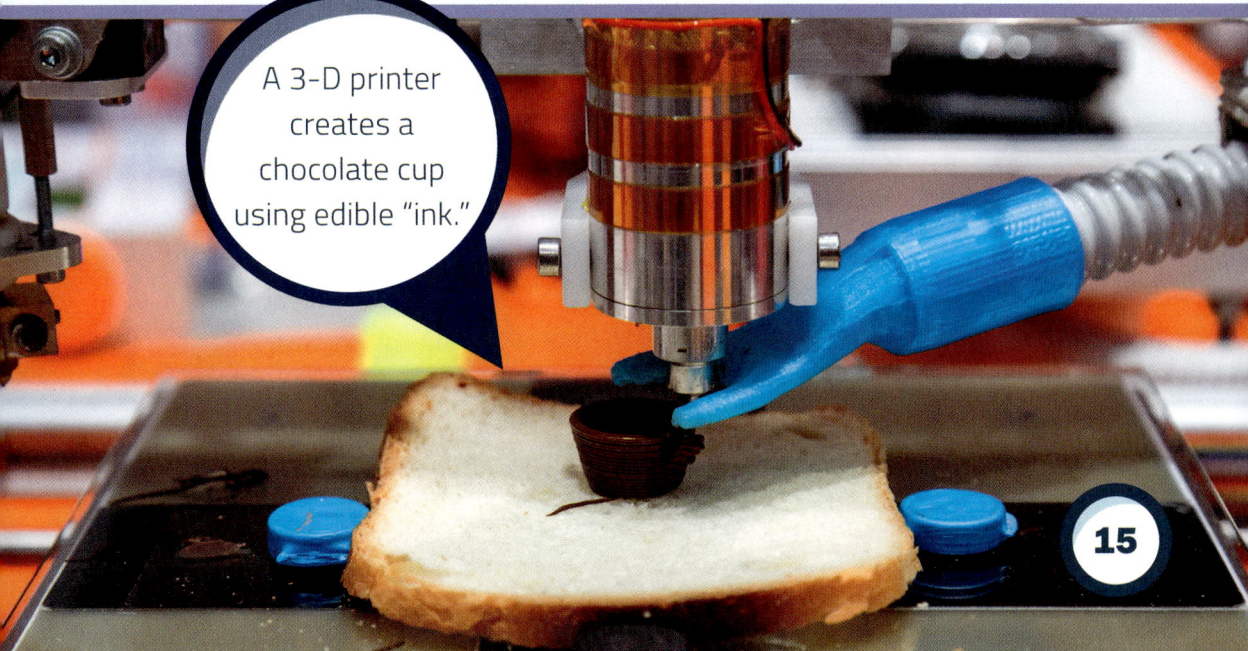

A 3-D printer creates a chocolate cup using edible "ink."

Lab-Grown Organs Improve Transplants

Organ transplants are lifesaving medical procedures. But they are also complex. An organ must be a perfect match to a recipient's body. Many times, a body rejects a transplanted organ. Researchers have been searching for a way to reduce organ rejection.

One promising technology is lab-grown organs. Scientists can create new tissue using stem cells.

Stem cells are simple cells in the body. The tissue can be turned into more complex cell types, such as blood cells. Stem cells are inside every person's body. Organs made from someone's own stem cells are less likely to be rejected. Stem cells can be turned into cartilage cells. This can help repair a damaged knee. Researchers are developing techniques to grow organs from

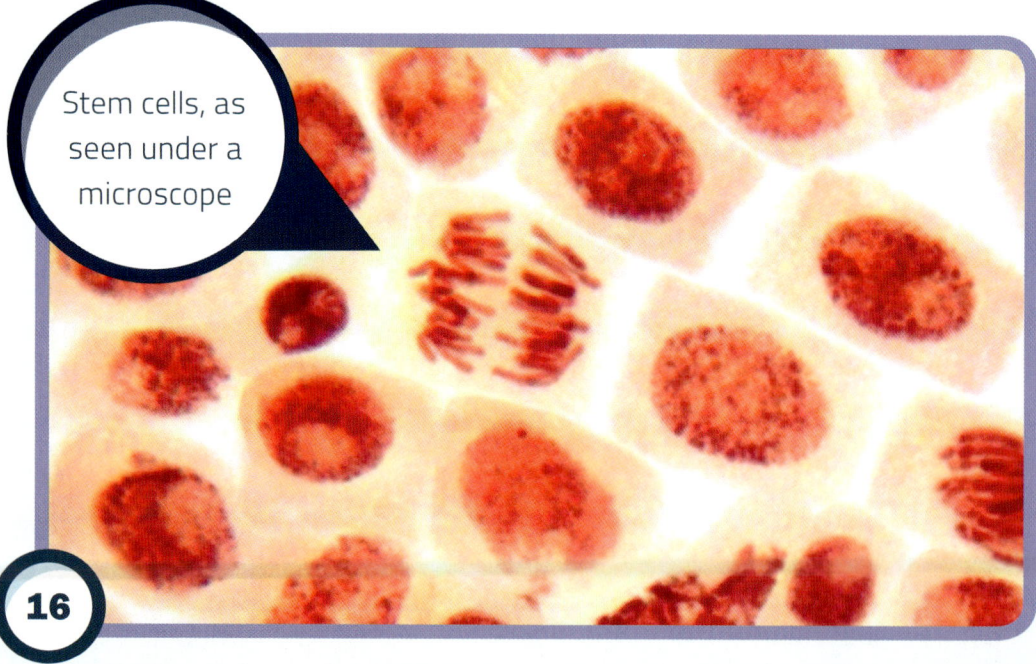

Stem cells, as seen under a microscope

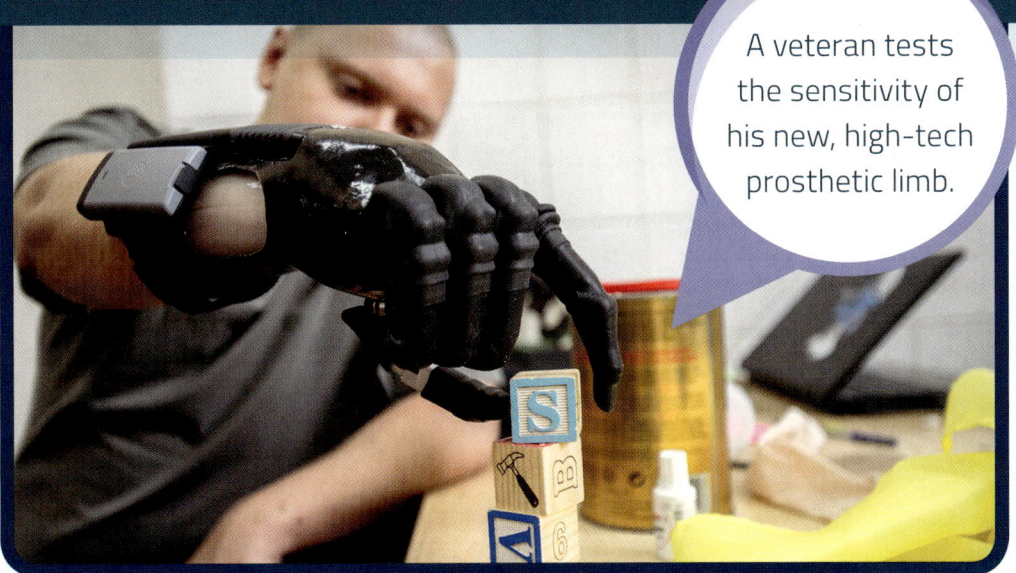

A veteran tests the sensitivity of his new, high-tech prosthetic limb.

stem cells. In the future, patients may get new hearts grown from their own stem cells.

Stem cells are just one part of this promising technology. Stem cells need something to grow inside of or around. Researchers are making disposable structures called scaffolds out of special plastics. The plastics dissolve as the organ or tissue grows. Just the new organ or tissue is left behind.

0.10
Diameter, in inches (0.25 cm), of an artificial stomach grown to measure bacteria in the gut.

- Recipients' bodies often reject transplanted organs.
- Stem cells from a recipient's own body may help transplants succeed.
- Scientists grow organs on disposable scaffolds.

HIGH-TECH PROSTHETIC LIMBS

Prosthetic limbs have existed for centuries. But today, they are getting some high-tech upgrades. Some prosthetic hands have finger joints. The joints move and work as real fingers do. Scientists are working on limbs that have the power of touch. Wearers could feel heat or even pain.

17

Maglev Trains Race into the Future

In the 1860s, Americans built a railroad network that stretched from coast to coast. Trains are not a common way to travel in the United States today. But they remain popular in other parts of the world. Engineers are developing ways to make trains faster. The fastest trains do not even touch the tracks.

These new trains use magnetic levitation, or maglev for short. Some magnets pull toward each other. This is called attraction. Others push away from each other. This is called repulsion. Maglev trains use magnetic attraction and repulsion to hover over the tracks.

A maglev train pulls into the station in Shanghai.

1,800

Estimated speed, in miles per hour (2,900 km/h), of the next generation maglev train.

- Maglev trains use magnetic laws of attraction and repulsion to hover over their tracks.
- This means the trains create much less friction than traditional trains do.
- This technology allows trains to reach speeds of up to 350 miles per hour (560 km/h).

MAGLEV TO SPACE

Maglev technology could be used for more than train travel. Some researchers are developing a maglev space shuttle. A tunnel of tracks would go up along the side of a mountain. A maglev shuttle carrying cargo would travel along the tracks. It would be shot through the tunnel at speeds close to 20,000 miles per hour (32,000 km/h). This would launch the shuttle into orbit.

Traditional trains roll over train tracks. When a train wheel moves over the track, it causes friction. This friction slows the train down. But maglev trains hover over their tracks. There is less friction. This means maglev trains can reach much higher speeds. Traditional trains reach only 150 miles per hour (240 km/h). But maglev trains whoosh down the tracks at speeds up to 350 miles per hour (560 km/h).

An experiment demonstrating magnetic levitation

Outlet-Free Electricity Changes How We Charge

Every gadget that runs on electricity needs to be charged. To charge phones, computers, or tablets, users need to find electrical outlets. But being plugged in means users are connected to one place.

Wireless charging pads help users ditch their charging cords. Inside the pad is a strong magnet. Wires wrap around the magnet. Electrical current flows through the wires. This creates an electromagnetic field. This field can pass the electrical current charge to the device on the pad. Some pads look like a coaster and sit discreetly on a table or counter. Some companies are building furniture with wireless chargers built in.

Cell phones do not need to plug into wireless chargers.

These chargers can help get rid of the tangle of cords many households have. But there is a catch. The wireless charger still has to be plugged into an outlet, even if the device does not. Wireless chargers may be a cool technology. But they still require a conventional electrical outlet to work.

A new technology may change that. These new charging stations would attach to the walls of a home or business. Like wireless charging pads, they would create an electromagnetic field. But the field would be much, much bigger. It could charge devices up to 7 feet (2.1 m) away. Users could charge multiple devices anywhere within the field.

700

Amount of power, in milliamp hours, a wireless charger from WattUp could provide.

- Electronic devices must be charged from an electrical outlet.
- Wireless charging pads do not require a physical connection to a device, but they still must be plugged in.
- New charging stations could create large electromagnetic fields, letting users charge devices from anywhere within those fields.

Personal-Assistant Robots Take Charge

Personal-assistant robots are available to purchase today. Devices such as the Amazon Echo or Samsung Otto are popular examples. These technologies talk to other electronics. They can turn on lights and check the weather. Some can even monitor a house for potential break-ins.

Echo and Otto are useful devices. But they look more like speakers than humanlike robots. Scientists are currently developing new robots that look more like humans. Humanlike robots could be more helpful. They could move around the house or office. They could do chores or run small errands. They could even provide entertainment. Imagine a robot reading a bedtime story. It could make the story

Pepper is a personal-assistant robot available for sale in Japan.

20
Number of engines that power Pepper, a personal-assistant robot developed in Japan.

- Personal-assistant robots already exist, but they don't look like humans.
- Companies are developing robots that act more like humans.
- Such robots would need to read emotions and have serious conversations.

ROBOT PETS

Researchers believe robot pets could be more useful. Robot pets could become companions for the elderly. One company has created robot cats that interact with their owners. The robots purr and nuzzle. They act like real cats. But unlike real cats, robot cats do not need to be fed. They do not have to visit the vet or use the litter box. But they may have to be repaired periodically.

come to life with voices and lighting effects.

A personal-assistant robot would have to be very intelligent. It would need to recognize and remember faces. It would need to tell a person's mood. Robots could do more than search the web. They could hold real conversations. Over time, they could become members of the family.

Robot pets may become more helpful in the future.

23

Self-Driving Cars Take to the Roads

Experts predict that by 2026, few people will drive cars anymore. But there will still be millions of cars on the road. How is this possible? Self-driving technology.

Companies have been testing self-driving car technology for years. Self-driving vehicles must be able to react to the world around them.

They must know when to brake. They have to change lanes safely. Most importantly, they must react to avoid accidents.

Humans are fairly good at these driving skills. But self-driving technology testing has found something surprising. Self-driving cars may be *better* at some driving

24

The interior of self-driving cars may look very different from cars today.

A driverless small bus was put to the test in Spain in 2014.

skills than humans are. Computers aboard the cars identify objects zipping past on the road. These computers are correct 95 percent of the time. That is better than how human drivers perform on this task.

The more self-driving cars on the road, the safer the roads should become. This is because the self-driving cars can communicate with one another. They always know when another car is close. This allows them to keep a safe distance away from one another.

3.9

Minimum distance, in inches (10 cm), within which self-driving cars can pinpoint obstacles.

- Experts predict self-driving cars will be common by 2026.
- Self-driving cars cause fewer accidents and could improve traffic.
- Computers in self-driving cars are better at identifying things in their surroundings than humans are.

THINK ABOUT IT

Would you buy a self-driving car? Why or why not? Use evidence from this chapter to support your answer.

Virtual Reality Builds Whole New Worlds

Virtual reality (VR) used to be science fiction. Characters on *Star Trek* could step into a special room. There, computers could create an entire virtual world. Such adventures seemed too futuristic ever to be real. But today, VR is not science fiction. It is actual technology. And it has some amazing uses.

One promising use for VR is scientific discovery. In the Cave Automatic Virtual Environment (CAVE), scientists can interact with their research. An image is projected onto the walls and floor of a room. A scientist puts on a pair of 3-D glasses. Virtual projections pop up and seem to float in midair.

Virtual reality glasses are bulky, but can make users feel like they are in a different world.

Biologists could use CAVE to get inside a human body. They could walk inside a giant human heart. This could help them see the organ from the inside out. Medical students could also use CAVE for learning. They could use VR to practice surgery.

A user does not need a big room to experience VR. Special VR goggles can be purchased in stores today. Oculus Rift went on sale in the spring of 2016. Users put on these goggles to play interactive games. They can even explore what it is like to be an astronaut in space.

12

Height, in feet (3.7 m), of the largest CAVE room the Visbox company makes.

- Virtual reality has gone from science fiction to new technology.
- Researchers can explore their research objects in interactive 3-D in a CAVE virtual reality room.
- Users can play games using VR goggles.

NASA scientists use CAVE to test a space mission they developed.

Fact Sheet

- In the future, 3-D printers may make it possible for anyone with one of these devices to design and print a prosthetic device for someone who has lost a limb. The e-NABLE network brings together engineers, physicians, families, and amputees to design prosthetics that are more functional and comfortable.

- A growing world population puts a lot of stress on technology. The United Nations expects that by 2050, close to 10 billion people will live on Earth. This will increase the demand for food by 70 percent compared to 2016. To meet this demand, the international organization recommends reducing food waste and producing more food with less water. It predicts scientists will develop new ways to irrigate fields and reduce crop loss.

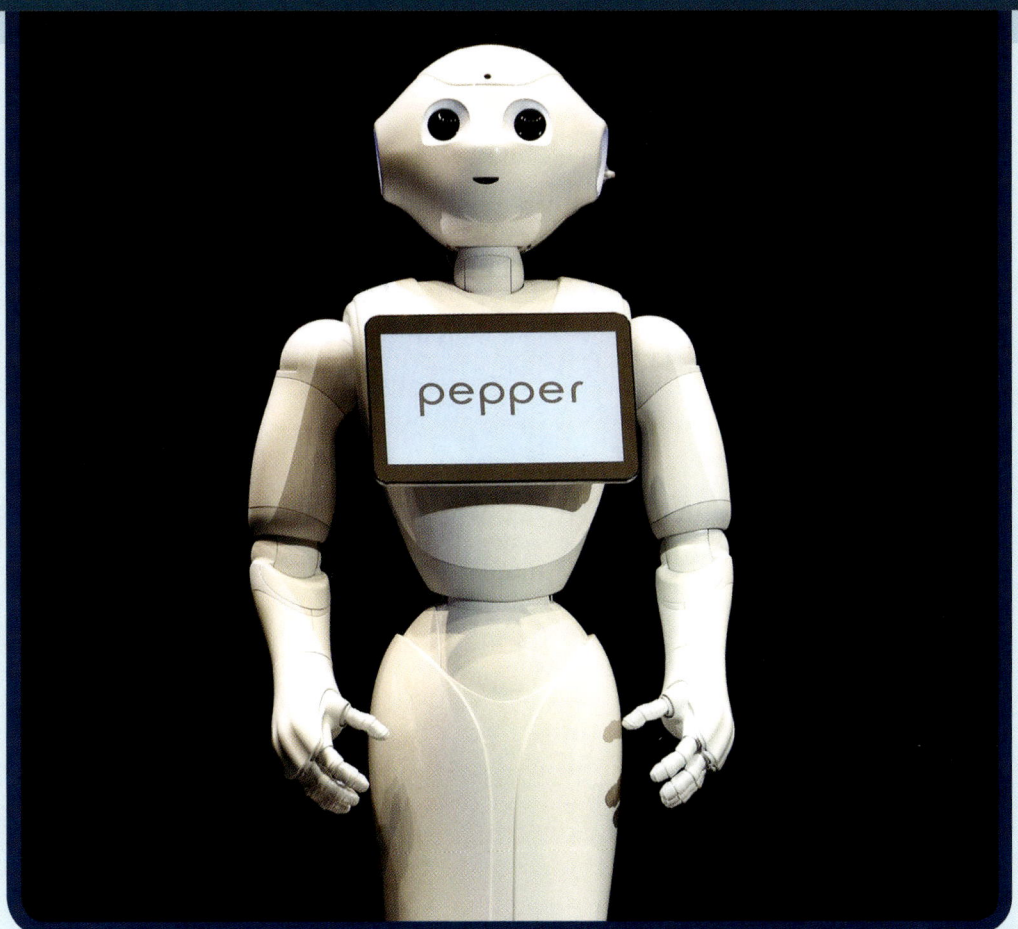

- Augmented and virtual reality both change the way people interact with the world. VR will help scientists interact with their research subjects. The Microsoft HoloLens helps students in medical school see inside the human body without the use of a scalpel. Instead, students wear VR headsets that allow them to see and manipulate a virtual model of the human body and its systems.

- Technology will improve the transportation industry, too. Telematics technologies record people's driving habits. They can tell drivers how long they have been driving, if they accelerate or brake too quickly, or if they are speeding. Drivers can use the data to make them safer on the road.

Glossary

archaeologist
Someone who studies human civilization by digging up historical sites.

exchange
To give one type of money and receive another type of money worth the same amount.

friction
Resistance or slowing down that occurs when two objects rub together.

implanted
Inserted into the body.

membrane
A barrier through which some objects can flow but others cannot.

molecules
The smallest particles of a substance that still have all the qualities of that substance.

population
All the people who live in a town, city, state, country, or world.

prosthetic
An artificial body part, such as a hand, arm, or leg.

transplant
Organs that are taken from one body and put into another.

For More Information

Books

Bedell, Jane. *So, You Want to Be a Coder? The Ultimate Guide to a Career in Programming, Video Game Creation, Robotics, and More!* New York: Aladdin, 2016.

Ceceri, Kathy. *Robotics: Discover the Science and Technology of the Future with 20 Projects.* White River Junction, VT: Nomad, 2012.

Mullenbach, Cheryl. *The Industrial Revolution for Kids: The People and Technology That Changed the World: With 21 Activities.* Chicago: Chicago Review, 2014.

Visit 12StoryLibrary.com

Scan the code or use your school's login at **12StoryLibrary.com** for recent updates about this topic and a full digital version of this book. Enjoy free access to:

- Digital ebook
- Breaking news updates
- Live content feeds
- Videos, interactive maps, and graphics
- Additional web resources

Note to educators: Visit 12StoryLibrary.com/register to sign up for free premium website access. Enjoy live content plus a full digital version of every 12-Story Library book you own for every student at your school.

Index

About the Author

Meg Marquardt loves writing about science. She enjoys researching physics, geology, and climate science. She lives in Madison, Wisconsin, with her two scientist cats, Lagrange and Doppler.